The Masonry Glossary

International Masonry Institute

THE MASONRY GLOSSARY

CBI

CBI Publishing Company, Inc.
51 Sleeper Street
Boston, Massachusetts 02210

Production Editor: M. Patricia Cronin
Text/Cover Designer: Betsy Franklin
Illustrator: Arabesque Composition
Compositor: Williams Graphics

Library of Congress Cataloging in Publication Data

The Masonry Glossary.
1. Masonry—dictionaries. I. International
Masonry Institute.
TA670.M36   693'.k'0321
81–194   ISBN 0–8436–0134–5
AACR2

Copyright © 1981 by The International Masonry Institute
823 Fifteenth Street, N.W. Washington, D.C. 20005.

Printing *(last digit):* 9 8 7 6 5 4 3 2 1
Printed in the United States of America

FOREWORD

The International Masonry Institute hopes that *The Masonry Glossary* will fill a long and broadly felt need by craftsmen, contractors, builders, architects, engineers, and others for a single comprehensive reference source of masonry terms.

For *The Masonry Glossary*, IMI set out to compile all the terms that are currently used in masonry construction, and to define those terms simply and accurately. The Institute hopes that it has mostly succeeded, but it has no illusions that its success may be total: First, because masonry is thousands of years old and is world-wide in application, the number of masonry terms is very large. It must be assumed that a few will have been overlooked. Second, the meanings given to masonry terms vary from one region or area to another. We have tried to take such variables into account, giving what we believe is the most commonly used definition. But there most certainly will be some instances where the judgement of others will differ from ours.

Nevertheless, we believe that the *Glossary* will be of great interest and utility to all the many craftsmen, businessmen, and professionals who are involved, to one degree or another, with masonry.

IMI appreciates the assistance given by the following organizations in reviewing material included in the *Glossary*:

> International Union of Bricklayers and Allied Craftsmen
>
> Mason Contractors Association of America
>
> Masonry Institute of America
>
> Masonry Institute of Houston-Galveston
>
> Indiana Limestone Institute of America, Inc.

Illustrations in the *Glossary* are not always in scale, having in some cases been deliberately drawn out of scale for the sake of clarity. And in other cases some technical details have been omitted from the drawings for the same reason.

IMI

The Masonry Glossary

abacus. The uppermost member of the capital of a column; often a plain square slab.

abate. In stone carving or sand blasting clay masonry, to cut away material, leaving parts in relief.

absorption. The amount of water that a (solid or hollow, clay or concrete) unit absorbs, when immersed in either cold or boiling water for a stated length of time; expressed as a percentage of the weight of the dry unit.

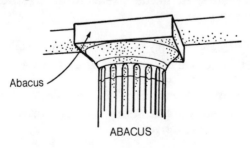

Abacus

ABACUS

1

abutment. The supporting wall or pier that receives the thrust of an arch.

accelerator. In masonry, any ingredient added to mortar to speed drying and solidifying. (Compare **retarding agent.**)

admixtures. Materials that are added to mortar or grout to impart special properties to the mortar or grout.

adobe brick. Large, roughly molded, sun dried clay bricks of varying sizes.

aggregate. In terrazzo work, a granule, other than marble, used in toppings (i.e., abrasives, quartz, granite, river gravel, synthetic types, etc.). In general, marble is referred to as "chips."

alabaster. Fine-grained, translucent variety of gypsum; generally white or delicately shaded. (This term is also incorrectly applied to fine-grained marble.)

anchor. Metal rod, wire, or strap that secures building veneer stone or other masonry to structural framework, backup wall, or other elements, or holds stone units together. (See **cramp.**)

angle closer. A portion of a whole brick that is used to close the bond of brickwork at corners.

angle iron. A structural iron angle; used for lintels to support masonry over openings, such as doors or windows.

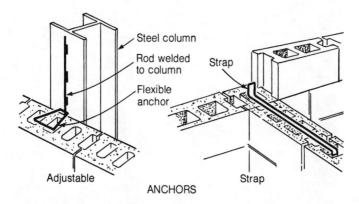

Adjustable

Strap

ANCHORS

apex stone. Uppermost stone in a gable, pediment, vault, or dome.

arch. A curved compressive structural member, spanning openings or recesses; also built flat.

back arch. A concealed arch carrying the backing of a wall where the exterior facing is carried by a lintel.

jack arch. One having horizontal or nearly horizontal upper and lower surfaces. Also called a *flat* or *straight arch.*

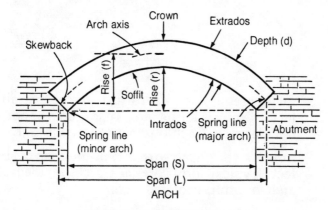

ARCH

THE MASONRY GLOSSARY

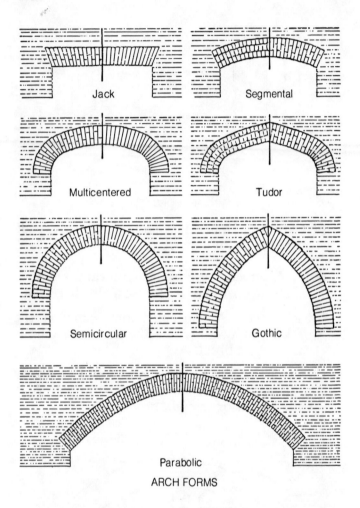

Jack

Segmental

Multicentered

Tudor

Semicircular

Gothic

Parabolic

ARCH FORMS

major arch.   Arch with spans greater than six feet. Typical forms are Tudor arch, semicircular arch, Gothic arch, or parabolic arch.

minor arch.   Arch with maximum span of six feet. Typical forms are jack arch, segmental arch, or multi-centered arch.

International Masonry Institute

relieving arch. An arch built over a lintel, flat arch, or smaller arch to divert loads, thus relieving the lower member from excessive loading. Also known as a *discharging* or *safety* arch.

trimmer arch. An arch, usually a low rise arch of brick, used for supporting a fireplace hearth.

architrave. Lowermost unit of an entablature, carried by columns (or their capitals) or pilasters.

argillite. Metamorphic rock resulting from the hardening of siltstone and/or claystone and shale.

arkose. Sandstone containing feldspar grains in abundance.

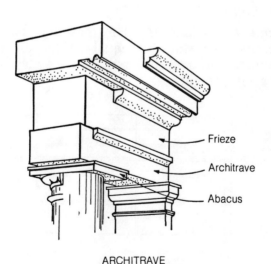

Frieze

Architrave

Abacus

ARCHITRAVE

THE MASONRY GLOSSARY

arris. External angular intersection between two planar faces, or two curved faces (as in moldings), or between two flutes on a Doric column, or between a flute and the fillet on an Ionic or a Corinthian column.

artificial stone. A contradiction in terms, as stone is a naturally occurring earth material. This phrase is used to describe materials variously called art marble, artificial marble, cast stone, terazzo, patent stone, and reconstructed stone. A mixture of stone chips or fragments embedded in a matrix of cement or plaster with the surface ground, polished, molded, or otherwise treated to simulate stone.

ashlar masonry. Masonry composed of variable size rectangular units having sawed, dressed, or squared bed surfaces, properly bonded, and laid in mortar.

coursed ashlar. Ashlar masonry laid in courses of stone of equal height for each course, although different courses may be of varying height.

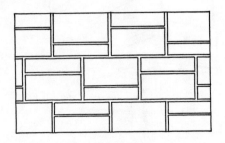

COURSED ASHLAR

International Masonry Institute

random ashlar. Stone masonry pattern of rectangular stones set without continuous joints and laid up without drawn patterns. If composed of material cut to modular heights, discontinuous but aligned horizontal joints are discernible.

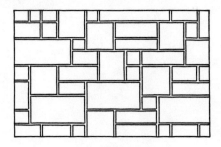

RANDOM ASHLAR

B

back filling.   (1) Rough masonry built behind a facing or between two faces. (2) Filling over the extrados of an arch. (3) Brickwork in spaces between structural timbers, sometimes called *brick nogging.*

backup (or backing).   The part of a multi-wythe masonry wall behind the exterior facing.

baluster.   A small column, which is generally turned that supports the rail of a balustrade.

balustrade.   An ornamental fencing consisting of a series of balusters supporting a handrail or molding.

band course.   See **belt course.**

banker.   Bench of timber or stone (may be a single block) on which stone is shaped.

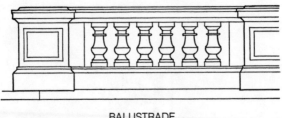

BALUSTRADE

bar tracery.   See **tracery**.

barge stone.   A generally projecting masonry unit that edges the slope of a roof or makes up the defining edges of a gable wall. (This term is a corruption of an earlier term *verge stone.*)

basalt.   Dark, fine-grained igneous rock used extensively as a paving stone and occasionally as a building stone.

bat.   A piece of brick, usually half the full size or smaller.

batted work.   Hand-dressed stone surface scored with a batting tool from top to bottom in narrow parallel strokes. Strokes may be vertical (in which case the surface may be called *tooled*) or oblique. There are six to ten strokes per inch. Batting is also called *broad tooling, droving,* or *angle dunting.*

batter.   Masonry that is receding or sloping back in successive courses; the opposite of a *corbel.*

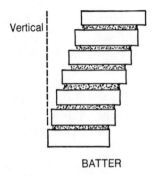

Vertical

BATTER

batting tool. A mason's chisel, several inches wide, used to dress stone to a striated surface.

bearing partition (or **bearing wall**). A wall that supports a vertical load in addition to its own weight.

bed. (1) A layer (stratum) of rock between two bedding planes. (2) In layered stone used for building, a surface parallel to the stratification. (3) In construction, the bottom surface of the masonry unit as it lies in the wall or other structure.

bed joint. The horizontal layer of mortar on which a masonry unit is set.

bedding plane. The surface at which two beds, layers, or strata of rock join.

Belgian block. A type of paving stone generally cut to truncated, pyramidal shape and set with the base of the pyramid down.

belt course. A continuous, horizontal band of masonry marking a division in the wall plan. Sometimes called *band course, string course,* or *sill course.*

bevel. One side of a solid body that is inclined with respect to the other, with the angle between the two sides being other than a right angle.

blind tracery. A carved wall ornamentation in low relief using elaborate flowing or geometrical patterns.

block. In quarrying, the large piece of stone, generally squared, that is taken from the quarry to the mill for sawing, slabbing, and further fabrication. Also, in concrete masonry, a usually hollow unit. (See **concrete masonry unit.**)

blocking. A method of bonding two adjoining or intersecting walls, which were not built at the same time, by means of offsets in which the vertical dimensions are not less than eight inches.

bluestone. A trade term applied to hard, fine-grained, commonly feldsparic and sicaceous sandstone or siltstone, of a dark greenish- to bluish-gray color, which splits readily along bedding planes. This splitting forms the thin slabs commonly used to pave surfaces for pedestrian traffic. A variety of flagstone.

boasted work. Hand-dressed stone surface showing roughly parallel narrow chisel grooves. The grooves are not uniform in width and are not carried across the face of the stone.

bond. (1) In masonry, the arrangement of units to provide strength, stability, and pattern. (2) Cohesion force between mortar or grout and masonry units or reinforcement.

American bond. A form of bond in which every sixth course is a header course and the intervening courses are stretcher courses.

basketweave bond. Module groups of brick laid at right angles to those adjacent.

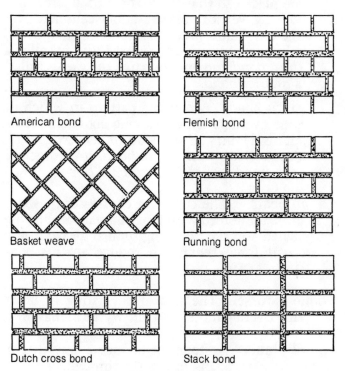

American bond

Flemish bond

Basket weave

Running bond

Dutch cross bond

Stack bond

BOND PATTERNS

THE MASONRY GLOSSARY

Dutch cross bond. A bond having the courses made up alternately of headers and stretchers. Same as an *English cross bond.*

Flemish bond. A bond consisting of headers and stretchers alternating in every course and laid so that they always break the joint. Each header is placed in the middle of the stretchers in the courses above and below.

random bond. Masonry constructed without a regular pattern.

running bond. Units in successive courses are placed so that the vertical head joints lap. Placing vertical mortar joints centered over the unit below is called a *center* or *half bond,* while lapping one-third of the way is called a *third bond* and one-fourth of the way is called a *quarter bond.*

stack bond. A bonding pattern in which no unit overlaps either the one above or below: all head joints form a continuous vertical line. Also called a *plumb joint bond, straight stack, jack bond,* and a *checkerboard bond.*

bond beam. The course or courses of masonry units reinforced with longitudinal bars and designed to take the longitudinal flexural and tensile forces that may be induced in a masonry wall.

bond course. The course consisting of units that overlap more than one wythe of masonry.

bond stone. A stone set in such a way that it carries through, or nearly through, a thick masonry wall to tie the wall together. The long dimension is generally perpendicular to the wall; but a very large bond stone may be set with its long dimension parallel to the wall and still serve as a bonder.

bonder. A masonry unit that overlaps two or more adjacent wythes of masonry to bind or tie them together. Also called a *bond header.* (see **header.**)

bonding agent. In terrazzo work, materials generally applied to "thinset" terrazo (i.e., latex, epoxy, polyurethane, or other types of adhesives). Used to increase adherence of the terrazzo mix to an existing base slab.

boss. (1) In masonry, a roughly shaped stone set to project for carving in place. (2) Carved ornamentation to conceal the jointing at the junction of ribs in a Gothic vault.

bossage. In masonry, the collective term for bosses that are placed together in a special area and ready for carving.

boulder. Naturally rounded rock fragment larger than 250 millimeters (10 inches) in diameter.

breaking joints. Any arrangement of masonry units that prevents continuous vertical joints from occurring in adjacent courses.

breccia. A rock characterized by coarse, angular fragments. The fragments are formed either by a crushing and natural recementing essentially in place or by the deposition of angular pieces that become consolidated.

brick. A solid masonry unit having a shape of approximately a rectangular prism. Most bricks are made from clay, shale, fire clay, or a mixture of these, and then fired; but brick may be composed of other materials, such as concrete brick and calcium-silicate brick.

acid resistant brick. Brick suitable for use in contact with chemicals, usually in conjunction with acid-resistant mortars.

adobe brick. Large, roughly molded, sun-dried or low-fired clay brick. Adobe bricks vary in size.

angle brick. Any brick shaped to an oblique angle to fit a salient corner.

arch brick. (1) Wedge-shaped brick for special use in an arch. (2) Extremely hard-burned brick from an arch of a scove kiln.

building brick. Brick for building purposes not especially treated for texture or color. Also called *common brick*.

clinker brick. A very hard-burned brick whose shape is distorted or bloated due to nearly complete vitrification.

common brick. See **building brick.**

dry press brick. Brick formed in molds under high pressures from relatively dry clay (five to seven percent moisture content).

economy brick. This brick's nominal dimensions are 4″ × 4″ × 8″.

engineered brick. This brick's nominal dimensions are 4″ × 3.2″ × 8″.

facing brick. Brick made especially for facing, or exposure purposes, and often treated to produce special surface textures. These bricks are made of selected clays, or treated, to produce the desired color.

fire brick. Brick made of refractory ceramic material that will resist high temperatures.

floor brick. Smooth dense brick, highly resistant to abrasion, used as finished floor surfaces.

gauged brick. (1) Brick that has been ground or otherwise produced according to accurate dimensions. (2) A tapered arch brick.

hollow brick. A masonry unit of clay or shale in which the net cross-sectional area in any plane parallel to the bearing surface is not less than 60 percent of its gross cross-sectional area measured in the same plane.

jumbo brick. A generic term indicating a brick larger in size than the standard. Some producers use this term to describe oversize brick of specific dimensions manufactured by them.

Norman brick. This brick's nominal dimensions are 4″ × 2⅔″ × 12″.

paving brick. Vitrified brick especially suitable for use in pavements where resistance to abrasion is important.

THE MASONRY GLOSSARY

Roman brick. This brick's nominal dimensions are 4″ × 2″ × 12″.

salmon brick. Generic term for underburned brick that is more porous and lighter colored than hard-burned brick. Usually pinkish-orange in color.

sewer brick. Low absorption, abrasive-resistant brick intended for use in drainage structures.

soft mud brick. Brick produced by molding (often by a hand process) relatively wet clay (20 to 30 percent moisture). When the insides of the molds are sanded to prevent the clay from sticking, the product is *sand-struck brick*. When the molds are wetted to prevent sticking, the product is *water-struck brick*.

stiff mud brick. Brick produced by extruding a stiff but plastic clay (12 to 15 percent moisture) through a die.

brick and brick. A method of laying brick so that units touch each other with only enough mortar to fill surface irregularities.

brick grade. A designation for the durability of the unit, expressed as *SW* for severe weathering, *MW* for moderate weathering, or *NW* for negligible weathering. Durability is determined by measuring compressive strength and absorption.

brick type. Designation for facing brick that indicates tolerance, chippage, and distortion. Expressed as *face brick standard* (FBS), *face brick extra* (FBX), and *face brick architectural*

(FBA) for solid brick, and *hollow brick standard* (HBS), *hollow brick extra* (HBX), *hollow brick architectural* (HBA), and *hollow brick basic* (HBB) for hollow brick.

broach. (1) In quarrying, to free stone blocks from the ledge by cutting out the webbing between holes drilled close together in a row. (2) To finish a stone surface with broad parallel grooves. A general term describing machine-worked stone finishes. Some broached work has a shallow drafted margin surrounding the broaching. (Compare **boasted work.**)

broad tooling. See **batted work**.

brownstone. A dark-brown and reddish-brown sandstone quarried and extensively used for building in the eastern United States during the middle and late Nineteenth Century. Most later use has been for renovation, repair, or additions to structures in which the stone was originally used.

brushed finish. Stone finish produced by a coarse, rotating wire brush.

building stone. Any stone that may be used in building construction; granite, limestone, marble, etc.

bull header. A header that is laid on its edge so that the end of the unit is exposed.

bull nose. A convex, semicircular molding used on edges of such stone units as stair treads, window sills, and partitions.

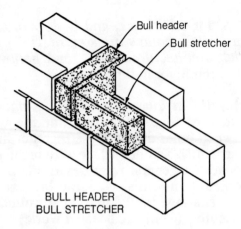

Bull header
Bull stretcher

BULL HEADER
BULL STRETCHER

bullnose block. A brick or concrete masonry unit having one or more rounded exterior corners.

bull stretcher. Any stretcher that is laid on its edge to show its broad face.

bushhammer. A hammer having a face that is sharply ridged or toothed with points in a square-set pattern.

bushhammer finish. A stone surface dressed with a bushhammer. Used decoratively or to

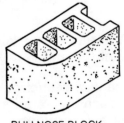

BULLNOSE BLOCK

International Masonry Institute

provide a roughened traction surface for treads, floors, and pavements.

butter. To place mortar on a masonry unit with a trowel.

buttress. A projecting mass of masonry set at an angle to or bonded into a wall that it strengthens or supports. A buttress decreases in its cross-sectional area from top to base.

button. In setting masonry, lead discs or other materials to carry the weight of the superincumbent stone while the mortar is green and curing.

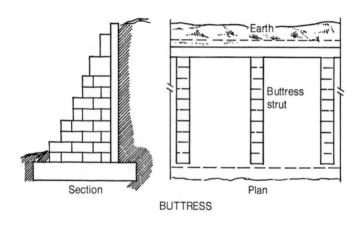

Section   Plan

BUTTRESS

calcite. A mineral form of calcium carbonate. Principal constituent of most limestones.

calcite streak. A former fracture or parting (in limestone) that has been recemented and annealed by the deposition of obscure white or light-colored calcite.

canopy. A shallow projecting roof, bracketed or cantilevered, ornamenting a doorway, window, niche, or throne.

cantilever. A structural member, supported at only one end, that projects from a support.

cap. Masonry units laid on top of a finished wall.

capacity insulation. The ability of masonry to store heat as a result of its mass, density, and specific heat.

capital. An intermediate member between the shaft of a column or pier and a beam, arch, or vault, usually ornamented by molding or carving or both.

capstone. Any single stone at the top of a masonry structure.

carved work. In stonework, hand cutting or ornamental features for which the lines cannot be applied from a pattern.

carver. In stone industry, the artisan who does carved work.

caryatid. A supporting member, serving the function of a pier, column, or pilaster, and carved or molded in the form of a draped human female figure.

cast stone. A precast building material manufactured from concrete.

cavity wall. A construction of masonry laid up with a continuous airspace between the wythes. The wythes are tied together with metal ties or bonding units (headers).

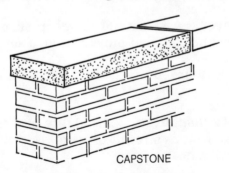

CAPSTONE

International Masonry Institute

c/b ratio. The ratio of the weight of water absorbed by a masonry unit when immersed 24 hours in cold water to the weight of water absorbed after an additional immersion for five hours in boiling water. An indication of the probable resistance of brick to freezing and thawing. Also called the *saturation coefficient.*

cell.   See **core.**

cement mortar.   A mixture of cement, lime, sand, or other aggregates and water used for plastering over masonry or to lay blocks.

centering.   Temporary formwork for the support of masonry arches or lintels during construction. Also called *center(s).*

ceramic color glaze.   An opaque colored glaze of a satin or gloss finish obtained by spraying the clay body with a compound of metallic oxides, chemicals, and clays. It is burned at high temperatures, fusing glaze to body and making them inseparable. See **clear ceramic glaze.**

chamfer.   To bevel an arris or edge.

chamfered rustication.   Rustication in which the smooth faces of the stones parallel to the wall are deeply beveled where they join to an internal angle of 135°, so that where the two stones meet the chamfering forms an internal right angle.

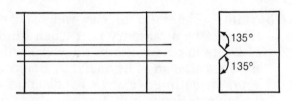

CHAMFERED RUSTICATION

chase.   A continuous recess in a wall to receive pipes, ducts, conduits, etc. The recess is usually vertical.

chase bonding.   Joining old masonry work to new by means of a bond having a continuous vertical recess the full height of the wall.

chimney.   A shaft built to carry off smoke.

chimney breast.   The projection of the interior or exterior face of a wall caused by fireplaces or flues.

chimney lining.   Fire clay or terra cotta material or refractory cement made to be built inside a chimney.

chimney throat.   That part of a chimney directly above the fireplace where the walls are brought close together.

chips.   Marble granules screened to various sizes.

class of unit.   A ranking of masonry units according to their different grades or types in ASTM specifications, the different raw materials

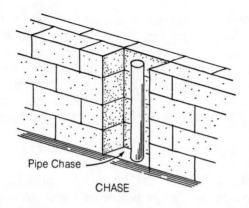

Pipe Chase

CHASE

they are manufactured from, or their different specified compressive strengths, tolerances, or other characteristics.

clay. A natural mineral aggregate consisting essentially of hydrous aluminum silicate. It is plastic when moistened, stiff when dried, and vitrified when fired beyond maturing temperature.

clay mortar mix. Finely ground clay used as a plasticizer for masonry mortars.

clean back. The visible end of a stone laid as a bond stone.

cleanout holes. Openings in the first course of masonry for cleaning mortar droppings prior to grout placement in grouted masonry. Required in high lift grouting.

clear ceramic glaze. Same as *ceramic color glaze* except that it is translucent or slightly tinted.

cleavage. In rocks, a tendency to split (cleave) along parallel and generally closely spaced surfaces caused by planar orientation of mineral constituents. True cleavage surfaces are unrelated to original stratification, but the term is also loosely used in some stone industries for splitting along the depositional layering.

clip. A portion of a brick cut to length.

closer. (1) The last masonry unit laid in a course. It may be whole or a portion of a unit. (2) A stone course running from one window sill to another (a variety of *string course*). See **angle closer, king closer,** and **queen closer**.

closure. Supplementary or short length units used at corners or jambs to maintain bond patterns.

cobble. Naturally rounded rock fragment between 60 millimeters (2½ inches) and 250 millimeters (10 inches) diameter. Used for rough paving, walls, and foundations.

collar joint. The vertical longitudinal joint between wythes of masonry.

color pigment. Inorganic matter used in the terrazzo mix to vary the color. A powdered substance that, when blended with a liquid vehicle, gives the matrix its coloring.

column. In structures, a relatively long, slender structural compression member such as a post, pillar, or strut. Usually vertical, a col-

umn supports a load that acts in the direction
of its longitudinal axis.

composite masonry. Multiple wythe construction in which at least one of the wythes is
dissimilar to the other wythe or wythes with
respect to type or grade of units or mortar.

concrete masonry unit. In masonry, a precast,
hollow block or solid brick of concrete used in
the construction of buildings.

"A" block. A hollow unit with one end
closed and the opposite end open, forming two
cells when laid in the wall.

bond beam block. A hollow unit with web
portions depressed 1¼ inches or more to form
a continuous channel, or channels, for reinforcing steel and grout. U-blocks are sometimes used to form bond beams, especially
over openings.

bullnose block. A concrete masonry unit
that has one or more rounded exterior corners.

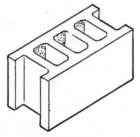

CONCRETE MASONRY UNIT

THE MASONRY GLOSSARY

cap block (or **paving unit**). A solid flat slab, usually 2¼ inches thick, used as a capping unit for parapet and garden walls. Also used for stepping stones, patios, veneering, etc.

channel block. A hollow unit with web portions depressed less than 1¼ inches to form a continuous channel for reinforcing steel and grout.

concrete block. A hollow concrete masonry unit made from portland cement and suitable aggregates such as sand, gravely crushed stone, bituminous or anthracite cinders, burned clay or shale, pumice, volcanic scoria, air-cooled or expanded blast furnace slags, with or without the inclusion of other materials.

concrete brick. A solid concrete masonry unit made from portland cement and suitable aggregates, with or without the inclusion of other materials.

coping block. A solid concrete masonry unit for use as the top and finished course in wall construction.

faced block. Concrete masonry units having a special ceramic, glazed, plastic, polished, or ground face or surface.

filler block. Concrete masonry unit for use in conjunction with concrete joists for concrete floor or roof construction.

flashing block. In masonry, metal flashing used to block a parapet wall and prevent roof leaks around such a wall.

H block. A hollow unit with both ends open and a continuous bond beam recess at the intersecting web.

header block. Concrete masonry units that have a portion of one side of the height removed to facilitate bonding with adjacent masonry such as brick facing.

jamb block. A block specially formed for the jamb of windows or doors, generally with a vertical slot to receive window frames, etc.

lintel block (or **U-block**). A masonry unit consisting of one core with one side open. (Usually placed with the open side up, like a trough, to form a continuous beam.)

offset block. A unit that is not rectangular in shape. Usually made as a corner block to keep the construction modular.

open end block. A hollow unit, with one end closed and the opposite end open, forming two cells when laid in the wall.

pilaster block. Concrete masonry units designed for use in the construction of plain or reinforced concrete masonry pilasters and columns.

return (or **L**) **corner block.** Concrete masonry unit designed for corner construction for walls of various thicknesses.

sash block. Concrete masonry unit that has an end slot to receive jambs of doors or windows.

scored block. Block with grooves that are in

a visual pattern. For example, the grooves may simulate raked joints.

shadow block. Block with a face formed in planes to develop surface patterns.

sill block. A solid concrete masonry unit used for the sills of openings.

single corner block. Concrete masonry unit that has one flat end. Used in the construction of an end or a corner of a wall.

slump block. Concrete masonry units (produced so that they "slump" or sag in irregular fashion before they harden) used in masonry wall construction.

solid block. A concrete masonry unit in which the net cross-sectional area in every plane parallel to the bearing surface is 75 percent or more of its gross cross-sectional area measured in the same plane.

split face block. Concrete masonry unit with one or more faces having a fractured surface. Used in masonry wall construction.

sculptured block. Block with specially formed surfaces, as a sculpturing block.

control joint. A groove that is formed, sawed, or tooled in a masonry structure to regulate the location and amount of cracking and separation resulting from the dimensional change of different parts of the structure, thereby avoiding the development of high stresses.

coping. The materials or masonry units used to form a cap or a finish on top of a wall, pier, chimney, or pilaster to protect the masonry below from water penetration. Commonly extended beyond the wall face and cut with a drip.

coquina. Coarse porous limestone composed of shells and shell fragments loosely cemented by calcite.

corbel. The projection of successive courses of masonry out from the face of the wall to increase the wall thickness or to form a shelf or ledge.

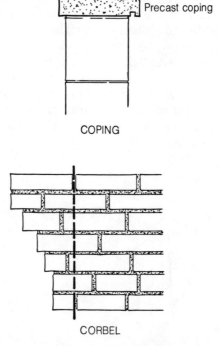

Precast coping

COPING

CORBEL

THE MASONRY GLOSSARY

core. A hollow space within a concrete masonry unit formed by the face shells and webs. The holes in clay units. Also called a *cell.*

cornerstone. (1) Generally a stone that forms a corner or angle in a structure. (2) More specifically, a stone prominently situated near the base of a corner in a building carrying information recording the dedicatory dates and other pertinent information. In some buildings, these stones contain or cap a vault in which contemporary memorabilia are preserved.

cornice. The molding or series of moldings forming the top member of a facade, door or window frame, or interior wall. Also the top member of a classical entablature.

corrosion resistant. (As it applies to anchoring or bonding material) corrosion resistant means that the material is treated or coated to retard harmful oxidation or other corrosive action. An example is steel galvanized after fabrication.

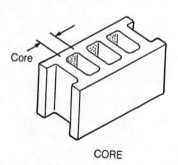

CORE

International Masonry Institute

course.   A layer (range) of masonry units running horizontally in a wall or, much less commonly, curved over an arch.

course bed.   Specially placed stone, brick, or other building material upon which other material is to be laid.

coursed ashlar.   See **ashlar masonry**.

coursed veneer.   In stone masonry, the use of veneer stones having equal height to form each continuous course. Horizontal joints extend the full length of any facade, but adjacent vertical joints are not superimposed.

cramp.   A U-shaped metal fastening to hold adjacent units of masonry together, as in a parapet or wall coping, or to secure marble slab veneers together.

cross-bedding.   In sedimentary rocks, inclined layers of sedimentation resulting from the progressive deposition of granular materials over a sloping surface. The layers are within a single bed between true bedding planes. Cross-bedding lends textural and color pattern to building stone.

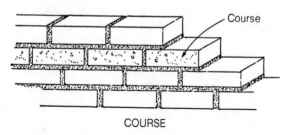

COURSE

THE MASONRY GLOSSARY

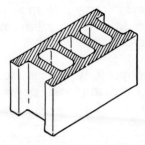

CROSS-SECTIONAL AREA

cross-sectional area. The net cross-sectional area of a masonry unit is the gross cross-sectional area minus the area of the cores or cellular spaces. The gross cross-sectional area of units is determined as the outside of the scoring, if any, but the cross-sectional area of the grooves formed by scoring is not deducted from the gross cross-sectional area.

crowfoot. See **stylolite**.

culls. Masonry units that do not meet the standards or specifications and that have been rejected.

curbing. Tabular bodies of stone or concrete, straight or curved, that are set on edge and form the upward vertical projection bordering streets, sidewalks, or planted areas.

curtain wall. A non-loadbearing exterior wall vertically supported only at its base, or having bearing support at prescribed vertical intervals.

International Masonry Institute

cut stone. Building stones cut to a specified size and shape. Each piece is fabricated to conform to plan drawings and to be installed in a designated location in the finished structure.

cutting stock. In stone milling, slabs of suitable size and thickness from which cut stone units are fabricated.

D

dab. To surface a stone with a pointed tool.

damp course. In stone masonry, an impervious horizontal layer to prevent vertical penetration of water in a wall. May be a course of tile or tight stone (e.g., slate or dense limestone), or a thin layer of asphaltic or bituminous material or metal. Damp courses are generally near grade to prevent upward migration by capillarity, but they are also used below copings, above roof level in chimneys, and elsewhere to stop downward seepage.

dampproofing. Prevention of moisture penetration due to capillary action by the addition of one or more coatings of a compound that is impervious to water.

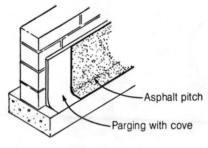

—Asphalt pitch

—Parging with cove

DAMPROOFING

dentil course. A narrow molding ornamented by small rectangular blocks (*dentils*) projecting at regular intervals.

diaper. Any continuous pattern in brickwork of which the various bonds are examples. It is usually applied, however, to diamond or other diagonal patterns.

dimension stone. Stone that is selected for or trimmed or cut to desired shapes and/or sizes. Used as building stone, markers, paving blocks or flagging, curbing, cut or carved ornaments and novelties, furniture (e.g., tabletops, laboratory bench tops and sinks), and industrial equipment that requires stone in shaped form (e.g., pebble mills or furnace liners).

dog's tooth. A brick laid with its corners projecting from the wall face.

dolomite. (1) Mineral form of calcium-magnesium carbonate. Constituent of some building limestones. (2) Limestone consisting principally of the mineral dolomite. Also called *dolostone*.

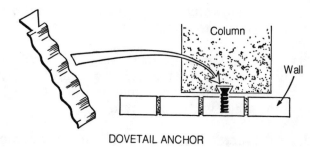

DOVETAIL ANCHOR

dolomitic lime. A trade term for high-magnesium lime. Also a misnomer as the product
does not contain dolomite.

dolomitic limestone. Limestone that contains
more than ten percent but less than eighty
percent of the mineral dolomite.

dovetail anchor. A splayed tenon that is
shaped like a dove's tail, that is broader at its
end than at its base, and that fits into the
recess of a corresponding mortise.

dowels. Straight metal bars used to connect two
sections of masonry.

drafted. Tooled border around the face of a stone
cut approximately to the width of the chisel.
Also called a *margin draft.*

dressed stone. Stone that has been worked to
its desired shape and that has had its exposed
face smoothed.

drip. Groove or slot cut beneath and slightly
behind the forward edge of a projecting stone
member, such as a sill, lintel, or coping, to

cause rainwater to drip off and prevent it from penetrating the wall.

dry. Natural fracture or parting (in stone) that has not been recemented or annealed by later deposition of mineral material. Contrasts (in limestone) with *glass seam.*

dry masonry. Masonry work laid without mortar.

dry mortar. A mortar in which the constituents are so proportioned that it is markedly stiffer than usual, yet has sufficient water for hydration.

dry pack. A mixture of portland cement and fine aggregate, dampened, but not to the extent that it will flow. It is usually rammed or packed in a hole to secure a bar or anchor, but is also packed under base plates.

dry wall. In masonry construction, a self-supporting rubble or ashlar wall laid up without mortar.

Dutchman. (1) A small piece of stone inserted as a filler in a patched area on a larger piece of dimension stone. (2) A small piece of stone inserted in an ashlar wall.

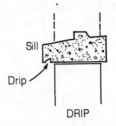

Sill

Drip

DRIP

International Masonry Institute

dwarf wall. A wall or partition that does not extend to the ceiling.

E

eccentricity. The distance between a vertical load reaction and a centroidal axis of masonry.

edgeset. A brick set on its narrow side instead of on its flat side.

edgestone. Stone used for curbing.

effective "b". The width of a wall that is assumed, in flexural computations, to work with reinforcing bars.

effective height. The height of a member that is assumed when calculating the slenderness ratio.

effective thickness. The thickness of a member that is assumed when calculating the slenderness ratio.

efflorescence.   A deposit or encrustation of soluble salts, generally white and most commonly consisting of calcium sulfate, that may form on the surface of stone, brick, concrete, or mortar when moisture moves through and evaporates on the masonry. Often caused by free alkalies leached from mortar, grout, adjacent concrete, or in clays.

empirical design.   A design based on the application of physical limitations learned from experience or observations gained through experience, without a structural analysis.

enclosure wall.   An exterior nonbearing wall in skeleton frame construction. It is anchored to columns, piers, or floors, but is not necessarily built between columns or piers nor wholly supported at each story.

engineered design.   A design based on a rational analysis, which takes into account the interrelationship of the various construction materials, their properties, and actual design loads, in lieu of empirical design procedures.

entablature.   In classical architecture, the elaborated beam member carried by the columns, horizontally divided into architrave (below), frieze, and cornice (above).

entasis.   Intentional slight convex curving of the side profiles in a tapered column to overcome the optical illusion of concavity that characterizes straight-sided columns.

International Masonry Institute

entasis

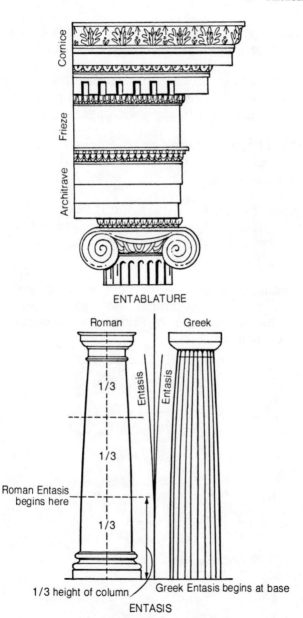

ENTABLATURE

ENTASIS

THE MASONRY GLOSSARY

epoxy joint. A visible joint filled with epoxy resin adhesive in place of mortar or caulking.

epoxy weld. In cut stone fabrication, a joint at an inside angle that is cemented by an epoxy resin to form an apparent single unit.

exfoliation. Peeling or scaling of stone or clay brick surfaces caused by chemical or physical weathering.

expansion bolt. An anchoring device (based on a *friction grip*) in which an expandable socket swells, causing a wedge action, as a bolt is tightened into it.

expansion joint. A vertical joint or space to allow for movement due to temperature changes (or other conditions) without rupture or damage.

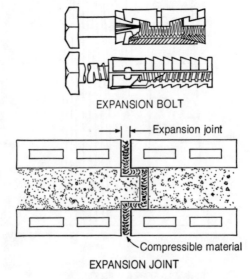

EXPANSION BOLT

EXPANSION JOINT

International Masonry Institute

face. (1) The exposed surface of a wall or masonry unit. (2) The surface of a unit designed to be exposed in the finished masonry.

face-bedded. Stone set with the stratification vertical.

faced wall. A wall in which the masonry facing and backing are bonded to act as a complete system under load.

face shell. The side wall of a hollow concrete masonry unit.

face shell bedding. Mortar is applied only to the horizontal face of the face shells of hollow masonry units.

facing. Any material, forming a part of a wall, used as a finished surface.

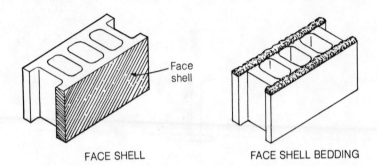

FACE SHELL   FACE SHELL BEDDING

false joint. A groove routed (and generally pointed) in a solid block of stone to simulate a joint.

fascia. A flat horizontal band that appears as a vertical face. The fascia is used decoratively, alone or in combination with other moldings.

fat mix. A mortar mixture containing a high ratio of binder to aggregate, thus providing better spread and workability.

feather-edged coping. Coping that slopes in only one direction (not ridged or gabled).

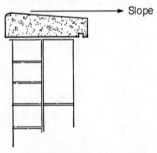

FEATHER-EDGED COPING

International Masonry Institute

field. The expanse of wall between openings, corners, etc., principally composed of stretchers.

fieldstone. Loose stone found on the surface or in the soil ("in the field"). Generally applied to slabby units, flat in the direction of the bedding or lineation of the rock and suitable for setting as dry wall masonry. Stream shingle has much the same shape and appearance, but is not found in the fields. Glacial or alluvial boulders and cobbles, which may be found in or on the soil, are not fieldstone in the strict sense.

filter block. A hollow, vitrified clay masonry unit, sometimes salt-glazed, designed for trickling filter floors in sewage disposal plants.

fire clay. A clay that is highly resistant to heat without deforming and used for making brick.

fireproofing. Any material or combination protecting structural members and increasing their fire resistance.

fire resistive material. See **noncombustible material.**

flagging. (1) Collective term for flagstones. (2) A surface paved with flagstones. (3) The process of setting flagstones.

flagstone. A flat stone, thin in relation to its surface area, that may be used as a stepping-stone, for a terrace or patio, or for floor paving. Usually either naturally thin or split from rock that cleaves readily.

flashing. (1) A thin impervious material placed in mortar joints and through air spaces in masonry to prevent water penetration and/or provide water drainage. (2) Manufacturing method to produce specific color tones in clay units.

flat arch (jack arch, straight arch). An arch that has little or no curvature.

flint. A dense, fine-grained, naturally occurring form of silica (SiO_2) that fractures conchoidally. A variety of *chert*, the more technical term. Most flint is gray, brown, black, or otherwise dark, but nodules and other chunks tend to weather white or change to lighter shades from the surface inward.

flush joint. See **joint.**

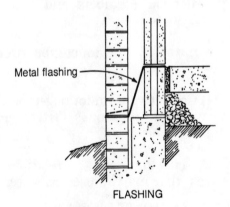

FLASHING

International Masonry Institute

freestone.   Stone with no tendency to split in any preferential direction, thus eminently suited for carving and elaborate milling. Restricted to stone that is fairly fine grained and works easily.

frieze.   (1) The middle member of a classical entablature. (2) A horizontal decorative band or border, carved or painted, encircling a room at dado level or higher, or used on exterior building faces.

frog.   A depression in the bed surface of a brick. Sometimes called a panel.

full mortar bedding.   Where mortar is applied to the entire horizontal face of the masonry unit.

furring.   A method of finishing the interior face of a masonry wall to provide space for insulation, to prevent moisture transmittance, or to provide a smooth or plane surface for finishing.

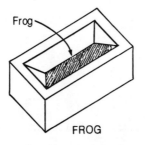

Frog

FROG

THE MASONRY GLOSSARY

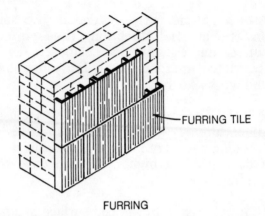

FURRING

furring units. Masonry units of shallow thickness used to finish the interior surface of a wall to provide space for insulation, to prevent moisture transmittance, or to provide a smooth or plane surface for finishing.

furrowing. The practice of striking a "V" in a bed of mortar with the point of the trowel.

International Masonry Institute

gallet. A stone chip or spall.

galleting. Insertion of chips or spalls of stone into the joints of rough masonry to solidify the wall, reduce the amount of mortar required, or add detail to the appearance.

gangsaw. A machine with multiple blades used to saw rough quarry blocks into slabs.

gargoyle. (1) A spout, commonly of stone but may be metal, tile, or other material, to discharge water outward from gutters, especially those behind parapets. (2) By usage, the carved or molded ornamentation, generally in the form of a grotesque figure, of a projecting gutter spout.

garretting. See **galleting**.

GARGOYLE

glass seam. Trade term (in limestone industry) for a former fracture or parting that has been recemented and annealed by the deposition of transparent calcite.

gneiss. Coarse-grained metamorphic rock with discontinuous foliation caused by planar alignment of platy and lath-shaped minerals.

granite. (1) In technical geologic terms, igneous rock with crystals or grains of visible size and consisting mainly of quartz and the sodium or potassium feldspars. (2) In building stone, crystalline silicate rock with visible grains. The commercial term thus includes *gneiss* (a metamorphic rock) and igneous rocks that are not granite in strict sense.

green mortar. Mortar that has set but not dried.

International Masonry Institute

greenstone. Metamorphic rock altered from basic (low-silica) igneous rock. The green color is due to iron-bearing silicate minerals. It is quarried and used as a structural and decorative dimension stone.

gross cross-sectional area. The total area parallel to the bearing surface of a masonry member or unit, calculated by using the overall actual dimensions of the member or unit.

ground. Nailing strips placed in masonry walls as a means of attaching trim or furring.

grout. A mixture of cementitious material and aggregate to which sufficient water is added to produce pouring consistency without segregation of the constituents.

grout lift. The height of which grout is placed in a cell, collar joint, or cavity without intermission.

grout pile. Colloquial term for stacked or piled quarried stone that cannot be further processed economically.

grout pour. The total grouted height between masonry lifts. A grout pour may consist of one or more grout lifts.

grouted cell masonry. Construction made with hollow units in which all cells and voids are filled with grout.

grouted masonry. Masonry construction made with solid masonry units in which the interior joints and voids are filled with grout.

high lift grouting. The technique of grouting masonry in lifts for the full height of the wall.

low lift grouting. The technique of grouting as the wall is constructed.

gypsum. Soft mineral consisting of hydrous calcium sulfate. The raw material from which plaster is made (by heating).

grog. Crushed brick that is blended with clay to form new brick.

hacking. (1) The procedure of stacking brick in a kiln or on a kiln car. (2) Laying brick with the bottom edge set in from the plane surface of the wall.

hard-burned. Nearly vitrified clay products that have been fired at high temperatures.

head. A stone that has one end dressed to match the face because the end will be exposed at a corner or in a reveal.

head joint. The vertical mortar joint between ends of masonry units. Also called a *cross joint* or a *vertical joint.*

header. A masonry unit that overlaps two or more adjacent wythes of masonry to tie them together. Also called a *bonder.*

59

blind header.   A concealed brick header in the interior of a wall, not showing on the faces.

bull header.   A header that is laid on its edge so that the end of the unit is exposed.

clipped header.   A bat placed to look like a header for purposes of establishing a pattern. Also called a *false header.*

flare flashed header.   A header of darker color than the field of the wall.

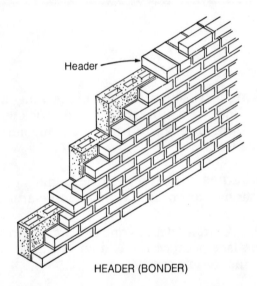

Header

HEADER (BONDER)

header course.   A continuous bonding course of header brick. Also called a *heading course.*

hearth.   (1) The masonry floor of a fireplace together with an adjacent area of fireproof material that may be a continuation of the flooring in the embrasure or some more decorative surfacing, as tile or marble. (2) An area permanently floored with fireproof material beneath and surrounding a stove.

height of wall.   The vertical distance from the foundation wall, or other similar intermediate support of such wall, to the top of the wall; or the vertical distance between intermediate supports.

herringbone work.   A pattern of setting in which the units in a wall are laid aslant, instead of flat, with the direction of incline reversing in alternate courses, forming a zigzag effect. In floors or paving, the units are set at approximately a 45° angle with the boundary of the area being clad, alternate rows reversing direction to give a zigzag horizontal pattern, and the unit in one row filling the triangle between two units in the adjacent row.

hewn stone.   Stone shaped with mallet and chisel.

high-calcium lime.   A lime that contains mostly calcium oxide or calcium hydroxide and not over five percent magnesium oxide or hydroxide.

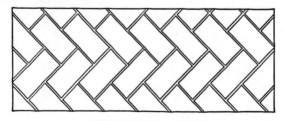

HERRINGBONE PATTERN

THE MASONRY GLOSSARY

high-magnesium lime. A lime produced by calcining dolomitic limestone or dolomite and thereby containing more magnesium oxide than limes made from calcite or high-calcium limestones and marbles. High-magnesium limes range from 37 to 41 percent MgO content, and high-calcium limes have less than 2.5 percent MgO. Also called (incorrectly) *dolomitic lime.*

hollow brick. A clay masonry unit in which the net cross-sectional area in every plane parallel to the bearing surface is not less than 60 percent of its gross cross-sectional area measured in the same plane.

hollow masonry units. A masonry unit in which the net cross-sectional area in any plane parallel to the bearing surface is less than 75 percent of its gross cross-sectional area measured in the same plane.

honed finish. In stone, a very smooth surface, just short of polished, imparted by a hand or mechanical rubbing process.

hung slating. (1) Slates covering a wall or other vertical surface rather than a roof (sloping) or floor (horizontal). (2) Slates supported by wire clips rather than by nails.

I

igneous rock. Rock formed by change of the molten material called magma to the solid state.

initial rate of absorption. The weight of water absorbed when a brick is partially immersed in water for one minute, expressed in grams per 30 square inches of contact surface. Also called *suction*.

initial set. The first setting action of mortar, the beginning of the set.

interlocking joint. A joint in which a rib or other protrusion on one stone complements a routed groove or slot on another to prevent relative displacement or movement.

intrados. The concave curve that bounds the lower side of the arch.

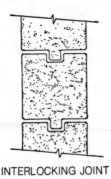

INTERLOCKING JOINT

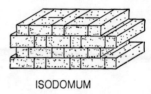

ISODOMUM

isodomum. An extremely regular masonry pattern in which stones of uniform length and uniform height are set so that each vertical joint is centered over the block beneath. Horizontal joints are continuous, and the vertical joints form discontinuous straight lines.

J

jack arch.   See **flat arch**.

jamb stone.   A stone constituting part of a vertical side in a wall aperture, such as a door or window opening.

joggle.   (1) An identation, projection, jog, or notch cut into or on a piece of building stone for fitting to a complementary offset in the adjacent stone in setting. (2) A piece of stone or metal that fits into paired apertures or grooves in two adjacent stones in a structure, keying them together. A dovetail joggle pre-

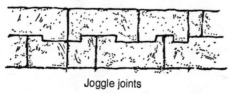

Joggle joints
JOGGLE

vents two stones from moving apart, whereas a simple joggle merely prevents lateral movement in one direction.

joint. The surface at which two members join or butt. If they are held together by mortar, the mortar-filled aperture is the joint.

joint reinforcement. Any type of steel reinforcement that is placed in or on mortar bed joints. Also called horizontal reinforcement.

jointing. The finishing of joints between courses of masonry units before the mortar has hardened.

jump. A step in a masonry foundation.

Weathered	Flush
Struck	Extruded
Concave	Raked
Deep Concave	V

JOINTING

International Masonry Institute

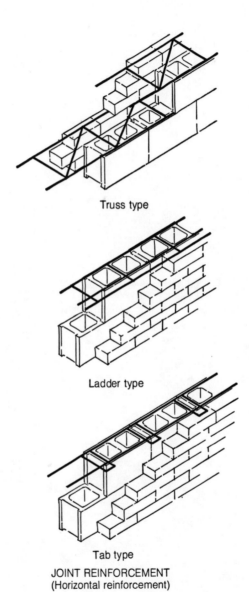

Truss type

Ladder type

Tab type

JOINT REINFORCEMENT
(Horizontal reinforcement)

THE MASONRY GLOSSARY

kerf. A cut or removal of material in a unit to facilitate breaking the unit to a desired shape or length.

key block. (1) The first block removed from a new quarry or ledge, providing space and access for further block removal by undercutting, underdrilling, or lateral shifting. (2) A keystone.

key course. (1) A horizontal row of keystones passing through the center of an arch. Generally used because the archway is too deep for a single keystone (or a single transverse row of arch stones) to suffice. (2) A course of keystones used in the crown of a barrel vault.

keystone. Wedge-shaped stone at the center or summit of an arch or vault, binding the structure actually or symbolically.

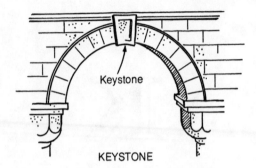

Keystone

KEYSTONE

kiln. A furnace, oven, or heated enclosure used for burning or firing brick or other clay material.

kiln run. Bricks from one kiln that have not yet been sorted or graded for size or color variations.

king closer. A brick cut diagonally to have one two-inch end and one full-width end.

kneeler. A building stone shaped to change a direction of the masonry, as (1) the stone that supports inclined coping on the slope of a gable, or that itself includes a length of coping

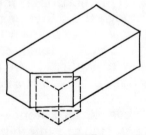

KING CLOSER

International Masonry Institute

(*skew table*) or (2) the stone that breaks the horizontal-vertical unit-and-joint pattern of the normal masonry wall to begin the curve or angle of an arch or vault.

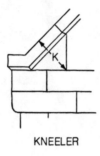

KNEELER

L

lap. The distance one masonry unit extends over another.

lapies. *Pronounced* lăp ē ĕź. The rugose bedrock surface formed beneath soil by the differential solution of limestone, gypsum, or other soluble rock. Generally deeply trenched along joints.

lateral support of walls. Method whereby walls are braced in the vertical span by beams, floors, or roofs, or in the horizontal span by columns, pilasters, buttresses, or cross walls.

laying overhand. Building the further face of a wall from a scaffold on the other side of the wall.

laying to bond. Laying the brick of the entire course without using a cut brick.

lead. The section of a wall built up and racked back on successive courses. A line is attached to leads as a guide for constructing a wall between them.

lewis. Any of several metal devices for lifting stone blocks in the quarry or mill or for hoisting columns or other heavy masonry units in construction.

box lewis. An assembly of metal components, some or all of which are tapered upward, that is inserted into a downward-flaring hole *(dovetail mortise)* and cut into the tops of columns or other heavy masonry units for hoisting.

lewis bolt. A bolt used to hang soffit stones or suspend the center part of lintels. May be conical or tapered and fit into slots cut in from the back. It may also be leaded into stone, or be combined with expansion sleeves. Carries the weight on an I-beam or other supporting member above.

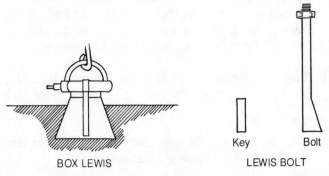

Key   Bolt

BOX LEWIS   LEWIS BOLT

International Masonry Institute

lewis hole. An opening that is cut or drilled in stone blocks, columns, or other heavy masonry units to receive lewis hoisting devices. The shape and size of the hole varies with the lewis that is to be used.

lewis pin. A metal peg, usually with its eye at the upper end, used for lifting stone blocks or masonry units. Lewis pins are used in pairs and are dependent on lever-action compression for gripping.

lime, hydrated. Quicklime to which sufficient water has been added to convert the oxides to hydroxides.

lime putty. Hydrated lime in plastic form ready for addition to mortar.

limestone. Rock of sedimentary origin composed principally of calcite or dolomite or both.

line. The string stretched taut from lead to lead as a guide for laying the top edge of a brick course.

line pin. A metal pin used to attach line used for alignment of masonry units.

liner. In the fabrication of stone veneer (principally marble), the stone bonded to the back of thin sheets to add strength, rigidity, bearing surface, or depth of joint.

lining. A wythe of similar masonry that is bonded to an existing wall to reinforce it.

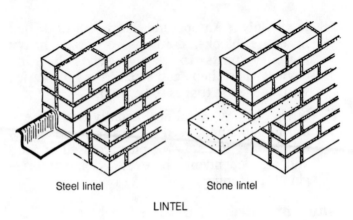

Steel lintel   Stone lintel

LINTEL

lintel. A beam placed or constructed over an opening in a wall to carry the superimposed load.

lintel course. A course of stone set at the level of a lintel, commonly differentiated from the wall by projecting, by finish, or by being lintel thickness, to continue the visual effect of the lintel.

loadbearing. A structural system or element designed to carry loads in addition to its own dead load.

lug. A projection from, or extension of, a building unit that engages an adjacent unit; for example, that part of a sill that extends into an adjoining jamb.

International Masonry Institute

lug sill.   A sill that projects into the jambs of a
window or door opening. (Compare **slip sill.**)

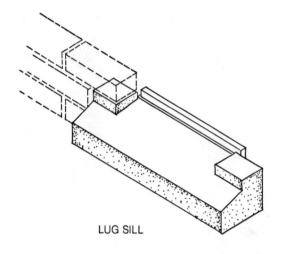

LUG SILL

machine finish. See **smooth machine finish**.

mall. See **mallet**.

mallet. A short-handled wooden hammer, with a truncated conical head, used to work stone and to drive mallet-head shaping tools.

marble. (1) In geology, a metamorphic rock made up largely of calcite or dolomite. (2) In dimension stone, a rock that will polish and that is composed mainly of calcite or dolomite or, rarely, serpentine.

    art marble. Artificial marble; precast terrazzo.

    broken marble. Fractured slabs of marble (not crushed by machines into chips).

margin draft. See **draft**.

mash hammer. A short-handled heavy hammer, with two round or octagonal faces, used to drive hammer-head shaping tools.

masonry. (1) Strictly speaking, the art of building in stone. By extension, masonry has come to mean the practice of the mason's craft with brick, tile, concrete masonry units, and other materials. (2) The work resulting from the practice of the mason's craft—structures built of stone, brick, or other materials set as units in patterns (and amenable to assembly with mortar, whether or not mortar is actually used). (3) The type of construction made up of masonry units laid up with mortar or grout or other accepted method of jointing.

plain masonry. Masonry constructed without steel reinforcement, except that which may be used for bonding or reducing the effects of dimensional changes due to variations in moisture content or temperature.

reinforced masonry. Masonry constructed with steel reinforcement embedded in such a manner that the two materials act together in resisting forces.

masonry cement. A mill-mixed cementitious material to which sand and water is added to make mortar.

masonry unit. Natural or manufactured building units of burned clay, concrete, stone, glass, gypsum, etc.

International Masonry Institute

hollow masonry unit. A unit in which the net cross-sectional area in any plane parallel to the bearing surface is more than 60 percent but less than 75 percent of the gross.

modular masonry unit. A unit in which the nominal dimensions are based on four inches or some multiple thereof.

solid masonry unit. A unit in which the net cross-sectional area in every plane parallel to the bearing surface is 75 percent or more of the gross.

mason's scaffold. Besides being totally self supporting, the true mason's scaffold must also carry the load of unusually heavy materials. It may be braced on a building already erected.

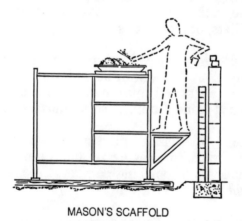

MASON'S SCAFFOLD

matrix. The portland cement and water mix or non-cementitious binder used to hold the marble chips in place for the terrazzo topping.

THE MASONRY GLOSSARY

metamorphic rock. Rock altered in appearance, density, and crystalline structure, and in some cases mineral composition, by high temperature or high pressure or both.

mica. A group of silicate minerals (muscovite and biotite are the most common) characterized by nearly perfect basal cleavage that causes them to split readily into extremely thin plates. The micas are prominent constituents of metamorphic and igneous rocks.

milling. A comprehensive term for the processing of quarry blocks through sawing, planing, turning, and cutting techniques to finished stone.

molding. The linear, continuous, decorative motif that is cut or carved on or into strips, billets, or blocks of stone.

monolithic. Shaped from a single block of stone, as a monolithic column, in contrast with a stacked column consisting of superimposed stone drums. A solid, massive unit.

mortar. A plastic mixture of cementitious materials, fine aggregate, and water. Generally made up of portland cement, lime, sand, and water.

fat mortar. Mortar containing a high percentage of cementitious components. It is a sticky mortar that adheres to a trowel.

high bond mortar. Mortar that develops higher bond strengths with masonry units than normally developed with conventional mortar.

lean mortar. Mortar that is deficient in cementitious components. It is usually harsh and difficult to spread.

mortar bed.   A thick layer of mortar used to seat a structural member.

mortar board.   A board about three feet square to hold mortar ready for the use of a bricklayer.

mosaic.   (1) A pattern or design formed by inlaying fragments or small pieces of stone, tile, glass, or enamel into a cement, mortar, or plaster matrix. (2) An irregular pattern of stone masonry surface used in veneering or solid stone walls.

MOSAIC

THE MASONRY GLOSSARY

N

natural bed. Setting stratified stone so that its bedding is horizontal (parallel to course joints). As tops and bottoms of beds are never recorded in quarrying, about half the stones are inverted, but they are not set on edge. This principle is occasionally applied to stone not visibly stratified; but to do so requires that the orientation be marked while the stone is being quarried, and then preserved until use.

natural cleft. Stone that is split (cleaved) parallel to its stratification, yielding an irregular but nearly flat surface.

natural stone. A redundancy, as stone is natural in its occurrence by definition. However the term is used to distinguish true stone from imitations.

net cross-sectional area. Average gross cross-sectional area of the masonry unit minus the area of ungrouted cores.

net section. Minimum cross section of the member under consideration.

nicked bit finish. A stone surface with parallel raised projections of various sizes and spacing, formed by an irregularly notched planer blade.

nidging (or **nigging**). A method of dressing stone, usually hard, by hand, using a pick or pointed hammer to furrow the entire surface.

nominal dimension. A dimension greater than a specified masonry dimension by the thickness of a mortar joint.

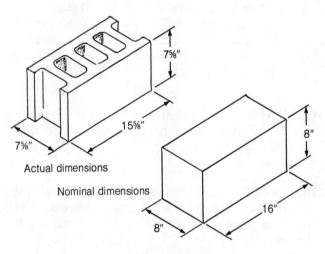

NOMINAL DIMENSIONS

International Masonry Institute

noncombustible.   Any material that will nei-
ther ignite nor actively support combustion in
air at a temperature of 1,200°F when exposed
to fire.

noncorroding.   Applies to anchoring or bonding
materials resistant to harmful oxidation or
other corrosive actions because of its composition (e.g., stainless steel, bronze, or copper).

nonstaining mortar.   A mortar with low free-
alkali content to avoid efflorescence to stain-
ing of adjacent stones by migration of soluble
materials.

offset. A course that sets in from the course directly under it; the opposite of a corbel.

offset block. A concrete masonry unit that is not rectangular. Usually used as a corner block to maintain the masonry pattern on the exposed face of a single-wythe wall whose thickness is less than half the length of the unit.

onyx. A banded, varicolored form of quartz.

oolite. (1) A spherical grain less than two millimeters in diameter that is most commonly composed of calcite and that consists of concentric shells. (2) A rock, generally limestone, composed largely of the spherical grains also called oolites.

oolitic limestone. Rock consisting mainly of calcite and made up largely or in considerable part of oolites or granular particles that may be tiny fossils or fossil fragments having oolitic coatings.

open slating. Pattern for installing slate shingles with spaces between adjacent slates in a course, providing ventilation if hung on open battens and reducing the amount of slate required. Spaces are covered by higher and lower courses. Also called *spaced slating*.

P

pacciarina. Residue from terrazzo grindings.

palletized. Material such as brick, block, or stone that is stacked on wooden platforms to permit mechanized handling.

panel. In terrazzo work, the space formed by the divider strips.

panel wall. A non-loadbearing exterior masonry wall having bearing support at each story.

parapet. A low wall around the perimeter of a building at roof level or around balconies.

parapet wall. The part of a wall that extends above the intersection of the wall with the roof.

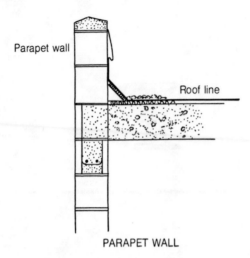

Parapet wall

Roof line

PARAPET WALL

parging. Plastering a coating of mortar, which may contain damp-proofing ingredients, over the back of masonry veneer, the face of the backup, or on underground exterior masonry.

parquetry. A flat pattern assembled of closely fitted pieces, usually geometrical. Many patterns consist of two or more colors or materials.

partition. An interior wall one story or less in height. It is generally non-loadbearing. In Canada, *partition* means an interior wall of one-story or part-story height that is never loadbearing.

patch. Compound used to fill natural voids or to replace chips and broken corners or edges in brick or in fabricated pieces of cut stone. Applied in plastic form. Mixed or selected to match the base material in color and texture.

International Masonry Institute

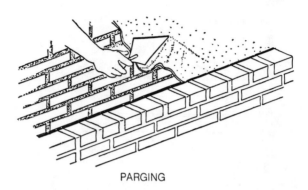

PARGING

patterned ashlar.   See **ashlar masonry.**

paver.   (1) A paving stone, brick, or quarry tile.
(2) A paving stone more than six inches
square.

paving stone.   A block or chunk of stone shaped
or selected by shape for surfacing a yard or
traffic surface.

pebble wall.   (1) Wall built of pebbles in mortar.
(2) Wall faced with pebbles embedded, at random or in pattern, in a mortar coating on the
exposed surface.

pediment.   The triangular face of a gable, if
separated by an entablature or molding from
the lower wall and treated as a decorative unit.
By extension, a triangular surface used orna-
mentally over doors or windows.

perpend stone.   A variety of bond stone that
extends completely through a masonry wall
and is exposed on both wall faces. A *through*
stone. Also *perpeyn wall.*

THE MASONRY GLOSSARY

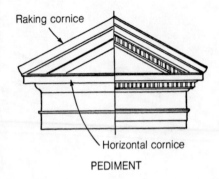

Raking cornice

Horizontal cornice

PEDIMENT

perpeyn wall. Wall built in the interior of a building and at a right angle to an enclosing wall, forming divisions like stalls, as in church buildings. Also *perpend wall.*

PERPEYN WALL

International Masonry Institute

phenocryst.   A course crystal in the fine-grained matrix of the igneous rock called porphyry.

picked finish.   A stone surface covered with small pits produced by a pick or chisel point striking the face perpendicularly.

pick and dip.   A method of laying brick whereby the bricklayer simultaneously picks up a brick with one hand and, with the other hand, enough mortar on a trowel to lay the brick. Sometimes called the *Eastern* or *New England method.*

pier.   An isolated column of masonry, or a bearing wall not bonded at the sides to associated masonry.

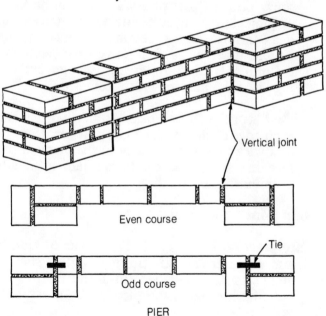

Vertical joint

Even course

Tie

Odd course

PIER

THE MASONRY GLOSSARY

pierced wall. A masonry wall in which an ornamental pierced effect is achieved by alternating rectangular or shaped blocks with open spaces.

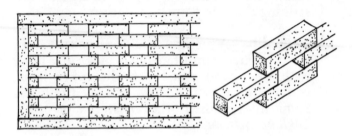

PIERCED WALL

pilaster. (1) A bonded or keyed column of masonry built as part of a wall. It may be flush or projected from either or both surfaces and has uniform cross section throughout its height. It serves as either a vertical beam or a column or both. (2) A flat engaged pier, extending less than half its width from a wall.

pitched stone. Rough-faced stone that has had each edge of the exposed face pitched. It is cut at a very low bevel, nearly in the plane of the face, in a straight line to form a defined arris at each mortar joint.

plain joint. See **joint**.

plate tracery. Tracery designs, usually simple and geometrical, cut through a thin slab of stone (as distinguished from *tracery* proper, which is formed by mortared sections of molding).

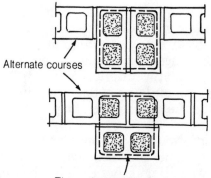

Alternate courses

Ties embedded in mortar joints

PILASTER

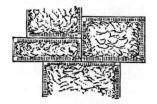

PITCHED STONE

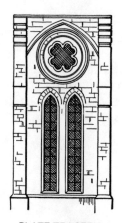

PLATE TRACERY

THE MASONRY GLOSSARY

plinth.   (1) A square or rectangular base for a column, pilaster, or door framing. (2) A monumental base, many of which are ornamented with moldings, bas reliefs, or inscriptions, to support a statue or memorial. (3) The base courses of a building collectively, if so treated as to give the appearance of a platform.

plucked finish.   Surface on stone produced by setting a planer blade so deep that it removes stone by spalling rather than by shaving.

plumb bob.   A shaped metal weight that is suspended from the lower end of a line to determine the vertical.

plumb rule.   A narrow board with parallel edges having a straight line drawn through the middle and a string attached at the upper end of the line for determining a vertical plane.

point.   A wedge-shaped or pyramidal chisel.

pointing.   (1) Troweling mortar into a joint after masonry units are laid. (2) Final treatment of joints in cut stonework. Mortar or a putty-like filler is forced into the joint after the stone is set. (3) In stone carving, creating points from a model and establishing their position on the stone that is to be carved.

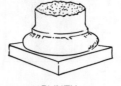

PLINTH

International Masonry Institute

polished finish. A finish so smooth that it forms a reflecting surface.

porphyry. Igneous rock characterized by two distinct and strongly contrasting sizes. Coarse crystals called phenocrysts are suspended in a finely crystalline groundmass or in one so fine grained that its crystallinity is invisible.

prefabricated masonry. Masonry fabricated in a location other than its final location in the structure. Also known as preassembled, panelized, and sectionalized masonry.

pressure-relieving joint. An open joint left at stated horizontal intervals to allow for expansion and contraction, commonly below horizontal supporting elements. Such joints are sealed with flexible caulking to prevent moisture penetration.

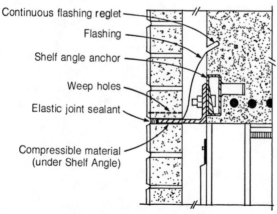

Continuous flashing reglet
Flashing
Shelf angle anchor
Weep holes
Elastic joint sealant
Compressible material
(under Shelf Angle)

PRESSURE-RELIEVING JOINT

THE MASONRY GLOSSARY

prism. A small assemblage made with masonry units and mortar and sometimes grout. Primarily used to predict the strength of full scale masonry members.

projection. A stone, brick, or block that has intentionally been set forward, at one end or throughout, of the general wall surface to appear more rugged, rustic, or to create a pattern.

quarry. An open excavation at the earth's surface for the purpose of extracting usable stone.

quarry run. Unselected materials of building stone within the ranges of color and texture available from the quarry that is the source.

quarry sap. Colloquial term for the natural moisture in stone as it comes from the quarry ledge. Varies in amount with the porosity.

quartzite. (1) Geologically, metamorphic rock resulting from the annealing of quartz sandstone. (2) In stone industry, a variety of sandstone composed largely of granular quartz and indurated either by metamorphism or cementation with silica to material that breaks with glassy fracture across grains and cementation alike.

quartzitic sandstone. Dimension stone trade term for a type of sandstone in which most of the grains are quartz and the cementing material is silica.

queen closer. A cut brick having a nominal two-inch horizontal face dimension.

quoin. (1) One of a series of masonry corner blocks, differing in size, finish, or material from the adjacent walling. (2) A wedge-shaped piece of stone. May be used in either the corner treatment described above (although most quoin stones are not wedge-shaped) or as a chock, a shim, or a device for leveling or aligning.

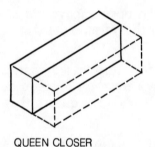

QUEEN CLOSER

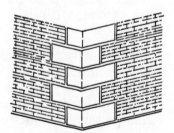

Stone quoins set in brickwork

QUOIN

International Masonry Institute

R

racking. Stepping back successive courses of masonry.

raggle. Slot or groove cut in masonry to receive mortared-in flashing.

raglet. A groove in a joint, or special unit, to receive roofing or flashing.

raglin. A joint raked in masonry to receive mortared-in flashing.

ragwork. Crude masonry laid up in a random pattern of thin-bedded undressed stone, like flagging. Commonly set mostly horizontal.

raked joint. See **joint**.

random ashlar. See **ashlar masonry**.

random bond. See **bond**.

random courses. Masonry set in courses of variable height.

random slates. Slate shingles installed in an irregular pattern using varying sizes.

range work. A course of any thickness that, once started, is continued across the entire face; but all courses need not be of the same thickness.

rebate. A rectangular groove or slot, as to receive a frame insert in a door or window opening. Also called a *sash groove*.

recess. A depth of some inches in the thickness of a wall such as a niche.

reconstructed stone. See **artificial stone**.

reglet. A recess to receive and secure metal flashing.

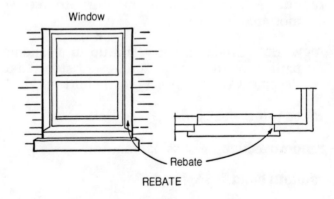

International Masonry Institute

regrating. Removing the surface of stone in place by some dressing method to expose fresh stone.

reinforced masonry. Masonry containing reinforcement in the grouted joints or grouted cores to resist shearing and tensile stresses.

relieved work. Ornamentation done in relief—that is, extending forward from a surface by shallow carving or molding.

relieving arch. An arch, usually blind, built into the wall above a lintel or flat arch to carry the load to walls or other supporting members.

repointing. Replacing mortar in masonry.

retarding agent. A chemical additive in mortar that slows setting or hardening, most commonly the SO_4—ion in the form of finely ground gypsum. (Compare **accelerator.**)

retempering. To moisten mortar and re-mix, after original mixing, to the proper consistency for use.

reticulated work. (1) Stone surface hand dressed to show a netlike or veinlike raised pattern. (2) A wall built of square blocks set diagonally, the joints showing a netlike pattern.

return. The shorter run in a right-angle turn that continues but changes the direction of a molding or wall.

THE MASONRY GLOSSARY

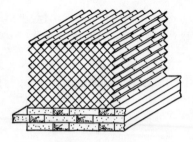

RETICULATED WORK

reveal. In the side of a door or window opening that is rebated for a frame, the surface extending from the slot (or frame) to the outer surface of the wall. (Compare **sconcheon**.)

revet. To face a sloping foundation or embankment with stone or concrete.

rift. Direction in which stone splits most readily. Term commonly used for granite or other stone without visible stratification or foliation.

riprap. Irregularly broken and random-sized large pieces of rock.

rise. The distance at the middle of an arch between the springing line and intrados or soffit.

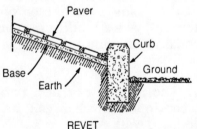

REVET

International Masonry Institute

rock. (1) Geologically, any natural mass of earth material that has appreciable extent. (2) In engineering, solid natural material that requires mechanical or explosive techniques for removal. (3) In the quarry industries, the term *stone* is more common and means firm, coherent, relatively hard earth material.

rock rash. A patchwork application of odd-shaped stone slabs, used on edge as a veneer, often further embellished with cobbles or geodes.

rodding. (1) Strengthening of stone slaps or panels by cementing steel reinforcing rods into routings in the back. Practice largely restricted to marble. (2) Slang for puddling or consolidating grout in a cavity or core.

roman arch. A semicircular arch. If built of stone, all units are wedge-shaped.

ROMAN ARCH

THE MASONRY GLOSSARY

rose window.   Large circular window, usually in a church facade, ornamented with tracery.

roughback.   (1) A concealed end of a stone laid as a bondstone. (2) Side cut (slab), having one side sawed and the other rough, from a block fed through a gangsaw.

rowlock.   A brick laid on its face edge with the end surface visible in the wall face. *Frequently spelled* rolok.

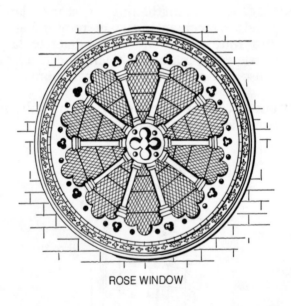

ROSE WINDOW

rubbed finish.   A stone finish between a smooth machine finish and a honed finish obtained by mechanical rubbing to a very smooth surface.

International Masonry Institute

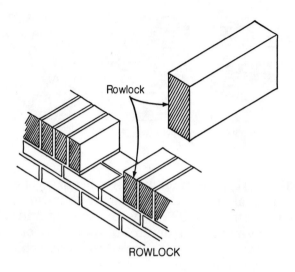

rubble. Pieces of broken stone, irregular in shape and size, used in the rough construction of walls, foundations, and paving. See also **snecked rubble**.

coursed rubble. Masonry composed of roughly shaped stones fitting on approximately level beds, well bonded, and brought at vertical intervals to continuous level beds or courses.

random rubble. Masonry composed of roughly shaped stones, well bonded and brought at irregular vertical intervals to discontinuous but approximately level beds or courses.

rough or ordinary rubble. Masonry composed of nonshaped field stones laid without regularity of coursing, but well bonded.

COURSE RUBBLE

RANDOM RUBBLE

ROUGH OR
ORDINARY RUBBLE

rustic. (1) A term describing masonry, generally of local stone, that is roughly hand dressed, and intentionally laid with high relief in relatively modest structures of rural character. (2) A grade of building limestone, characterized by coarse texture.

rustic joint. A deeply sunk mortar joint that has been emphasized by having the edges of the adjacent stones chamfered or recessed below the surface of the face.

rustic stone. A trade term for rough, broken stone suitable for rustic masonry. Generally set with the elongate dimension exposed horizontally. Most commonly limestone or sandstone, but can be any sound stone.

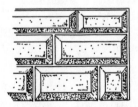
RUSTIC JOINT

International Masonry Institute

rusticated. Term describing cut stone walling with strongly emphasized recessed joints and smooth or roughly textured block faces. The border of each block may be rebated, chamfered, or beveled on all four sides, at top and bottom only, or on two adjacent sides. The face of the block may be flat, pitched, or diamond point, and if smooth may be hand or machine tooled.

S

S-iron. Generic term (because the "S" shape is so common) for exposed retaining plates on the ends of turnbuckled tie rods set between two masonry walls to prevent them from spreading or to secure an interior framing wall to a masonry wall. Star motif and other decorative shapes also used.

saddle. A strip of stone or wood used as a threshold.

saddle joint. A vertical joint along which the stone is lapped on either side to rise above the level of the wash on a coping or sill, thus diverting water from the joint.

saddle stone. See **apex stone**.

saddleback. A coping stone with its top surface shaped to wash (slope) in opposite directions, with the apex in the center of the width.

SADDLE BACK

sag. A depression in a horizontal line, meaning there is a slight fall below the level. Referring to a bricklayer's line.

salt glaze. A gloss finish obtained by a thermo-chemical reaction between silicates of clay and vapors of salt or chemicals.

sand-rubbed finish. The type of surface on a dimension stone obtained by rubbing with a sand-and-water mixture under a block. Commonly applied with a rotary or belt sander.

sand-sawed finish. The fairly smooth surface resulting from using sand as the abrasive agent carried by the gangsaw blades in stone fabrication.

sand-size. Grains between 1/16 millimeter (0.002 inch) and 2 millimeters (0.125 inch) in largest cross section.

sandstone. Sedimentary rock composed of sand-size grains naturally cemented by mineral material.

sawed face. See **sawed finish.**

sawed finish. Any stone surface left by a sawing process. The term is uninformative, but the names of the special sawed finishes, for example, sand-sawed and shot-sawed, are more used and more descriptive. Also called *sawed face.*

scabbing hammer. See **scabbling hammer**.

scabble. To dress stone to a rough planar face with a pick, scabbling hammer, or chisel, leaving prominent toolmarks.

scabbling hammer. A sledge with a pointed peen (pick) and a square head (less used) for the rough shaping of stone, particularly of blocks, at the quarry.

scagliola. Decorative inlay work in which mixtures of marble dust, a sizing, and various pigments are laid. Great care is taken with detail; and the figures, patterns, and designs are routed into a very flat surface on a plate or slab that is generally of marble, but may be other stone or plaster. The filled surface is then buffed to a polish, the designs are high-lighted and further tinted by delicate brushwork, and a transparent protective coating applied.

schist. Metamorphic rock with continuous foliation caused by the planar crystalline alignment of mica and other platy and lathlike minerals.

sconcheon. In the side of a door or window opening that is rebated for a frame, the strip extending from the slot (or frame) to the inner face of the wall. (Compare **reveal**.)

score. (1) To rout a channel or groove in stone finishing with hand tools or a circular saw to interrupt the visual effect of a surface or to otherwise decorate. (2) To roughen the surface of stone or concrete with straight gouges so that stucco or plaster will adhere.

scotia. One of the classical ornamental moldings, in profile showing a slightly asymmetrical concave curve.

sculpture. Statuary cut from stone by a sculptor using hand tools and polishing materials, with some assistance from powered cutting tools. The term is loosely used to describe statues modeled or cast rather than sculptured.

scutch. A mason's tool resembling a small pick used to trim units to a designed shape.

sectilia. A pavement made up of fitted hexagonal stones or tiles.

sedimentary rock. Rock formed from materials deposited as sediments in the sea, fresh water, or on land. The materials are transported to their site of deposition by running water, wind, moving ice, marine energy, or gravitational movements; and they may deposit as fragments or by precipitation from solution.

selenite. A variety of gypsum in transparent, foliated, crystalline form. Used as decorative building stone.

semi-rubbed finish. A surface of split stone sand rubbed to such a degree that the former prominences have been smoothed flat, but the lower areas still have a cleft surface.

serpentine. A group of minerals consisting of hydrous magnesium silicate, or rock largely composed of these minerals. Most commonly occurs in greenish shades, and is used for decorative stone, being the prominent constituent in some commercial marbles.

serpentine wall. A wall that is sine wave in plan.

SCOTIA

SERPENTINE WALL

THE MASONRY GLOSSARY

set. A change in mortar consistency from a plastic to a hard state.

setting space. The distance between the finished face of the wall and the backup wall as in masonry paneling or veneering.

shale. Clay that has been subjected to high pressures until it has hardened rock-like.

shot-sawed finish. The randomly scored surface in stone fabrication resulting from chilled steel shot carried by the gangsaw blades.

shoved joints. Vertical joints filled by shoving a unit against the next unit when it is being laid in a bed of mortar.

sill. A flat or slightly beveled stone set horizontally at the base of an opening in a wall.

sill course. A course set at a window-sill level and commonly differentiated from the wall by projecting, by finish, or by being sill thickness to continue the visual effect of the sill(s).

skew. In stone masonry, (1) a kneeler, and (2) in Scotland, a coping stone or the coping on a gable.

skew back. The incline surface on which the arch joins the supporting wall.

skew table. A variety of kneeler that is cut integrally with the lowest section of a gable coping and serves as a lower stop for sloping sections of the coping above.

slab. A broad, flat piece of stone cut or split from a block after quarrying. Especially used to mean the tabular sheet, ready for further fabrication, that comes from the gangsaw or wire saw.

slate. A hard, brittle metamorphic rock consisting mainly of clay minerals. It is characterized by good cleavage that is unrelated to the bedding in the earlier shape of clay from which it formed.

slenderness ratio. Ratio of the effective height of a member to its effective thickness.

slip sill. A stone sill set between the jambs of a window or door opening. (Compare **lug sill**.)

slushed joints. Vertical joints filled, after units are laid, by "throwing" mortar in with the edge of a trowel.

smooth machine finish (smooth planer finish, smooth finish). Stone surface obtained from a planer by use of a tool with a smooth edge set to shave without plucking. If toolmarks are evident, they may be removed by carborundum or other surfacing wheel, or by hand scraping.

sneck. Small, squared stone block used to fill interstices and to even out courses in rubble walls (hence *snecked rubble*).

snecked rubble.   Rubble masonry wall containing snecks.

soap.   A masonry unit of normal face dimensions, having a nominal two-inch thickness.

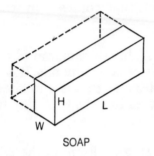

SOAP

soapstone.   Massive soft rock that contains a high proportion of talc and that is cut into dimension stone.

soffit.   The exposed lower surface of any overhead component of a building such as a lintel, vault, or cornice, or an arch or entablature.

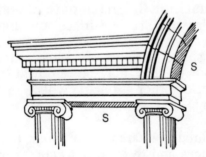

Soffit of an arch and of a lintel S

SOFFIT

International Masonry Institute

soft-burned. Clay products that have been fired at low temperature ranges, producing units of relatively high absorptions and low compressive strengths.

solar screen. A perforated wall used as a sunshade.

soldier. A stretcher set on end with its face showing on the wall surface.

solid masonry unit. A masonry unit whose net cross-sectional area in every plane parallel to the bearing surface is 75 percent or more of its gross cross-sectional area measured in the same plane.

spall. (1) (*verb, transitive*) To break away protrusions or edges on stone blocks with a sledge. (2) (*verb, intransitive*) To flake or split away through frost action or pressure. (3) (*noun*) A chip or flake.

spandrel. (1) A flat vertical face in an arcade bounded by the adjacent curves of two arches

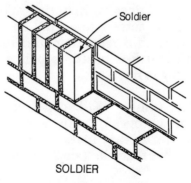

SOLDIER

THE MASONRY GLOSSARY

and the horizontal tangent of their crowns. When a lintel is used above an arched doorway or archway, two half-spandrels may sit astride the arch. (2) The facing of the area on buildings supported by a skeleton structure between the sill of one window and the top (or lintel) of the window next below.

splay. A reveal at an oblique angle to the exterior face of the wall.

split-face finish. A rough face formed by splitting slabs in a split-face machine.

split-face machine. A device that splits slabs of stone or concrete masonry units into usable thicknesses for job-fabricated masonry patterns. Generally hydraulic, but may operate on impact. Blades are used to split billets from slabs for most limestones and sandstones, but toothed bars may be used for harder stone such as granite.

springing line. The upper and inner edge of the line of skewback on an abutment.

spotting. Adhesive material, applied in plastic form and setting to a solid bonding area, that attaches thin veneering units to a backup wall and furnished multiple points at which the spacing is fixed.

squared rubble. Wall construction in which squared stones of various sizes are combined in patterns that make up courses as high as or higher than the tallest stones.

stack. Any structure or part thereof that contains a flue or flues for the discharge of gases. Also called a *chimney*.

stack bond. A bond pattern in which the masonry units are not lapped longitudinally in the face of the wall but are stacked vertically immediately over each other so as to form continuous joints both vertically and horizontally.

steatite. An industrial grade of talc that has high purity. Steatite block is soapstone that meets stated purity requirements.

stereobate. A basal pedestal-like structure or continuous basement wall supporting the higher parts of a classical building, but not carrying columns.

stereotomy. Cutting solids in three-dimensional shapes. The term is especially used to mean formal stone cutting by the rules of solid geometry, and by extension means the layout and design of such work and its placement in a structure.

sticking. Trade term used in the marble-fabricating industry for cementing together broken or separated stone.

stone. Rock selected or processed by shaping, cutting, or sizing for building or other use.

stone slate. Thin-bedded stone slabbing or flagging, irregular in size and shape, and generally limestone or sandstone, used as rough shingling on a roof. Unlike true slate, which is a metamorphic rock that splits along its cleavage, the stone slates separate along their bedding.

stonemason. A building craftsman skilled in constructing stone masonry.

stonework. (1) Masonry construction in stone. (2) Preparation or setting of stone for building or paving.

stool. Interior window sill, shelf, or ledge.

stop chamfer. A chamfer that curves or angles to become narrower until it meets the arris.

story. That portion of a building included between the upper surface of any floor and the upper surface of the next floor above, except that the top-most story shall be that portion of a building included between the upper surface of the top-most floor and roof above.

story pole. A marked pole for measuring masonry coursing during construction.

stratified rock. Layered earth materials deposited as successive beds of sediment and solidified by compaction, cementation, or crystallization.

stream shingle. Thin slabs of stone that accumulate in the channels of small high-gradient streams in a sloped, overlapping pattern resembling shingling—the pieces dipping

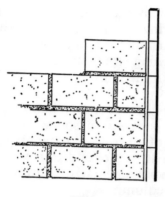

STORY POLE

upstream because they are most stable in that orientation. Much flagging and material termed fieldstone occurs as stream shingle. Only thin-bedded or foliated rocks form the flat pieces required, and limestone is the most common variety.

stretcher.   A masonry unit laid with its greatest dimension horizontal and its face parallel to the wall face.

bull stretcher.   Any stretcher that is laid on its edge to show its broad face.

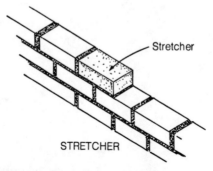

STRETCHER

THE MASONRY GLOSSARY

strike. To finish a mortar joint in stone setting or bricklaying with a stroke of the trowel, simultaneously removing extruding mortar and smoothing the surface of the mortar remaining in the joint.

string course (belt course, band course). A horizontal band of masonry, generally narrower than other courses, extending across the facade of a structure and in some structures encircling such decorative features as pillars or engaged columns.

stringing mortar. The procedure of spreading enough mortar on a bed to lay several masonry units.

strips. Billets of stone that are long in relation to the height of the exposed face.

struck joint. A joint from which excess mortar has been removed by a stroke of the trowel, leaving an approximate flush joint.

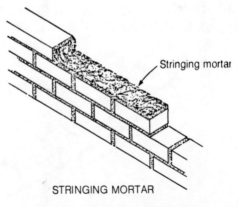

Stringing mortar

STRINGING MORTAR

International Masonry Institute

stylobate.   A large pedestal in the form of a basal structure or continuous basement wall that supports columns, or the stepped-back uppermost part of such a supporting structure.

stylolite. Colloquial: *crowfoot.* Generally a bedding plane in limestone and marble, and rarely a joint (fracture), along which differential solutions of the material on each side has caused interpenetration of points, cones, or columns to form a contact surface that is rough when separated. In cross section, the stylolitic surface has the appearance of a jagged, zigzag line of varying amplitude.

surfacing.   The grinding, grouting, and finishing operations on terrazzo topping.

T

tablet. (1) A stone or metal plate or bounded surface to carry words, letters, emblems, or carvings. (2) A coping stone set flat.

tailing in. Securing one end or edge of a projecting masonry unit, as a cornice.

talc. A soft mineral composed of hydrous magnesium silicate.

temper. To moisten and mix mortar to a proper consistency.

template. (1) The full-size sheet metal pattern to which a block or block face is cut. (2) Marble or other stone base for a toilet.

terrazzo. Marble-aggregate concrete that is cast in place or precast and ground smooth; used as a decorative surfacing on floors and walls.

tesserae. Thin slices of marble, colorful stone, or glass-like highly colored vitreous enamel material cut into squares or other shapes of any size. Used in mosaic work.

thinsets. Terrazzo systems that can be applied in a thin cross-section, ¾ of an inch or less.

throating. See **drip**.

through bond. The transverse bond formed by extending through the wall.

tie. Any unit of material that connects masonry to masonry or other materials. See **wall tie**.

tolerance. Specified allowance of variation from a size specification.

tooled finish. In stonework, a fluted, flat surface that carries two to twelve concave grooves per inch.

tooling. Compressing and shaping the face of a mortar joint with a special tool other than a trowel.

toothing. Constructing the temporary end of a wall with the end stretcher of every alternate course projecting. Projecting units are *toothers*.

topping. The wearing surface of the terrazzo floor.

International Masonry Institute

tracery. Gothic window ornamentation depending on window mullions in elaborate flowing or geometrical patterns built up of curved lengths of mortared stone molding.

traditional masonry. See **empirical design**.

transformed section. An assumed section of one material having the same elastic properties as the section of two materials.

travertine. A variety of limestone deposited by hot or cold water as cavern fillings, including stalactites and stalagmites, or as accumulations at springs.

tread. (1) The horizontal component of a stairstep. (Compare *riser*.) (2) The fore-to-aft dimension of a stairstep. (3) The upper surface of a step.

trig. The bricks laid in the middle of the wall that act as a guide to eliminate the sag in the line and to reduce the effect of wind blowing the line out of plumb.

trim. In building stone, that stone used as decorative members on a structure built or faced largely with other masonry material such as brick, tile, block, or terra-cotta. Trim items include sills, jambs, lintels, coping, cornices, quoins, and others. Also called *trimstone*.

tuck pointing.   The filling in with fresh mortar of cut-out or defective mortar joints in masonry.

tuff.   Rock composed of volcanic particles, ranging from ash size to small pebble size, compacted or cemented or welded to firm, consolidated state.

turned work.   In stone cutting, pieces with circular outline such as columns, balusters, and some bases and capitals. Generally cut on a lathe, although spheres and some other shapes may be cut by hand.

underbed.   Subsurface to accept terrazzo strips.

undercut.   (1) To cut away a lower part, leaving a projection above that serves the function of a drip. (2) To rout a groove or channel (a drip) back from the edge of an overhanging member.

V-cutting. Inscribed lettering in which the cuts are acutely triangular.

veneer. A single facing wythe of masonry units or similar materials securely attached to a wall for the purpose of providing ornamentation, protection, insulation, etc., but not so bonded or attached as to be considered as exerting common reaction under load.

verde antique. Dark-green serpentine rock marked with white veins of calcite.

vermiculated work. Stone surface incised with wandering discontinuous grooves resembling worm borings.

vertical joint. See **head joint.**

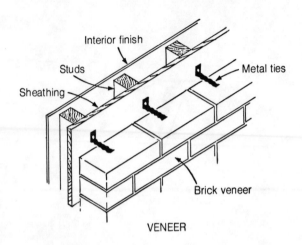

VENEER

vitrification. The condition resulting when kiln temperatures are so high as to fuse grains and close pores of a clay product, making the mass impervious.

vug. A pocketlike natural cavity in stone, generally the result of solution or recrystallization. Size not limited, but most are between a small fraction of an inch and a few inches in average diameter. May be lined with crystals or botryoidal layers of mineral materials.

wall. A vertical member of a structure, enclosing or dividing space.

apron wall. That portion of a wall below the sill of a window and above the floor.

area wall. (1) The masonry surrounding or partly surrounding an area. (2) The retaining wall around basement windows below grade.

bearing wall. See **loadbearing wall.**

cavity wall. A wall built of masonry units arranged to provide a continuous air space within the wall (with or without insulating material) and in which the inner and outer wythes of the wall are tied together with metal ties or headers.

composite wall. A multi-wythe wall in which at least one of the wythes is dissimilar to the other wythe or wythes with respect to type of masonry unit.

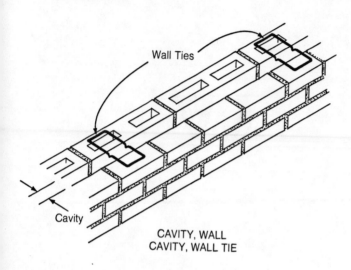

Wall Ties

Cavity

CAVITY, WALL
CAVITY, WALL TIE

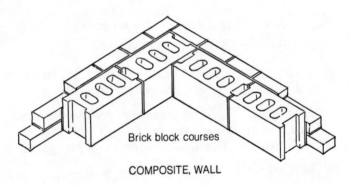

Brick block courses

COMPOSITE, WALL

curtain wall. An exterior non-loadbearing wall in skeleton frame construction. Such walls may be anchored to columns, spandrel beams or floors, but not necessarily built between columns.

dwarf wall. A wall or partition that does not extend to the ceiling.

International Masonry Institute

enclosure wall. An exterior nonbearing wall in skeleton frame construction. It is anchored to columns, piers, or floors, but not necessarily built between columns or piers nor wholly supported at each story.

exterior wall. Any outside wall or vertical enclosure of a building other than a party wall.

faced wall. A composite wall in which the masonry facing and the backing are so bonded as to exert a common reaction under load.

fire wall. Any wall that subdivides a building so as to prevent the spread of fire and that extends continuously from the foundation through the roof.

fire division wall. Any wall that subdivides a building so as to prevent the spread of fire.

foundation wall. A wall below the floor nearest grade serving as a support for a wall, pier, column, or other structural part of a building.

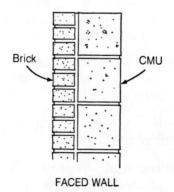

FACED WALL

THE MASONRY GLOSSARY

hollow masonry wall. A wall built of hollow masonry units.

hollow wall. A wall built of masonry arranged so as to provide an air space within the wall between the inner and outer wythes. (See also **cavity wall** and **masonry bonded hollow wall**.)

loadbearing wall. A wall that supports vertical load in addition to its own weight.

masonry bonded hollow wall. A hollow wall in which the facing and backing are tied together with solid masonry units.

non-loadbearing wall. A wall that supports no vertical load other than its own weight.

panel wall. An exterior non-loadbearing wall in skeleton frame construction, wholly supported at each story.

parapet wall. That part of any wall entirely above the roof.

party wall. A wall on an interior lot line, or any wall used or adapted for joint service between two buildings.

perforated wall. A wall that contains a considerable number of relatively small openings. Often called a *pierced wall.*

shear wall. A wall that resists horizontal forces applied in the plane of the wall.

single wythe wall. A wall of only one masonry unit in wall thickness.

solid masonry wall. A wall built of solid masonry units, laid contiguously, with the joints between the units filled with mortar or grout.

spandrel wall. That portion of a panel or curtain wall above the head of a window or door in one story and below the sill of the window in the story above.

veneered wall. A composite wall having a facing of masonry units or other weather-resisting noncombustible materials securely attached to the backing, but not so bonded as to intentionally exert common action under the load.

wall plate. A horizontal member anchored to a masonry wall to which other structural elements may be attached. Also called a *head plate.*

wall plug. Metal insert used for nailing wood furring and studs to masonry walls. Also called whistle anchor, nail clips.

wall tie. A bonder or metal piece that connects wythes of masonry to each other or to other materials.

cavity wall tie. A rigid, corrosion-resistant metal tie that bonds two wythes of a cavity wall.

veneer wall tie. A strip or piece of metal used to tie a facing veneer to the backing.

THE MASONRY GLOSSARY

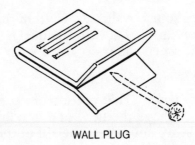

WALL PLUG

warped. Said of thin-bedded rock, such as flagging, having a natural curved or rippled finish similar to warped wood.

wash. A sloping upper surface of a building member, as a coping or sill, to carry away water.

wasting. In stone cutting, splitting off the surplus stone with a wedge-shaped or pyramidal chisel called a point, or with a pick. By either type of chisel, the faces of the stone are reduced to nearly plane surfaces, and the stone is said to be wasted off. In Scotland this is called *clowring*.

WASH

water retentivity. That property of mortar which prevents the rapid loss of water to masonry units of high suction. It prevents bleeding or water gain when mortar is in contact with relatively impervious units.

water table.   A projection of lower masonry on the outside of the wall slightly above the ground. Often a damp course is placed at the level of the water table to prevent upward penetration of ground water. Generally near grade and having a beveled top and a drip cut in the projecting underside to deflect water.

waterproofing.   Prevention of moisture flow through masonry.

waxing.   Trade term for filling cavities in finished marble for interior use with materials patterned and colored to match the piece.

web.   The cross wall connecting the face shells of a hollow concrete masonry unit.

weep hole.   An opening left (or cut) to prevent water from accumulating behind a wall or parapet or within a wall.

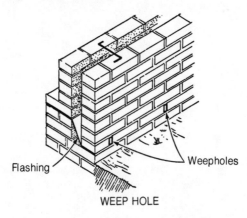

Flashing   Weepholes

WEEP HOLE

THE MASONRY GLOSSARY

wheel window. A circular window, usually large, having radial mullions of stone molding. Also called a *Catherine wheel window*, after the saint's symbol of martyrdom.

wire saw. An assembly for sawing stone, both in the quarry and in the mill, by a rapidly moving continuous wire (under tension and commonly helical) that carries a slurry or sand or other abrasive material through a slot that is deepened in the process.

wythe. A masonry wall, one masonry unit, a minimum of two inches thick.

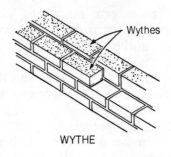

Wythes

WYTHE